Editor
Lorin Klistoff, M.A.

Managing Editor
Karen Goldfluss

Editor-in-Chief
Sharon Coan, M.S. Ed.

Cover Artist
Barb Lorseyedi

Art Director
CJae Froshay

Art Coordinator
Kevin Barnes

Imaging
Alfred Lau
James Edward Grace
Rosa C. See

Product Manager
Phil Garcia

Publishers
Rachelle Cracchiolo, M.S. Ed.
Mary Dupuy Smith, M.S. Ed.

W9-BTK-975

Word Problems

Practice Makes Perfect

GRADE 2

Authors

Teacher Created Materials Staff

Teacher Created Materials, Inc.
6421 Industry Way
Westminster, CA 92683
www.teachercreated.com

ISBN-0-7439-3312-5

©2002 Teacher Created Materials, Inc.
Reprinted, 2003
Made in U.S.A.

Table of Contents

Introduction

The old adage "practice makes perfect" can really hold true for your child and his or her education. The more practice and exposure your child has with concepts being taught in school, the more success he or she is likely to find. For many parents, knowing how to help their children can be frustrating because the resources may not be readily available. As a parent it is also difficult to know where to focus your efforts so that the extra practice your child receives at home supports what he or she is learning in school.

This book has been designed to help parents and teachers reinforce basic skills with their children. *Practice Makes Perfect* reviews basic math skills for children in grade 2. The math focus is word problems. While it would be impossible to include in this book all concepts taught in grade 2, the following basic objectives are reinforced through practice exercises. These objectives support math standards established on a district, state, or national level. (Refer to the Table of Contents for the specific objectives of each practice page.)

- adding and subtracting to 10
- adding and subtracting to 14
- adding and subtracting to 18
- adding three numbers
- adding and subtracting 2-digit numbers without regrouping
- adding and subtracting 2-digit numbers with regrouping

- using multiplication
- using and comparing fractions
- using time and money
- finding greater than and less than
- using a number line
- finding important information
- identifying place value

There are 36 practice pages organized sequentially, so children can build their knowledge from more basic skills to higher-level math skills. To correct the practice pages in this book, use the answer key provided on pages 47 and 48. Six practice tests follow the practice pages. These provide children with multiple-choice test items to help prepare them for standardized tests administered in schools. As children complete a problem, they fill in the correct letter among the answer choices. An optional "bubble-in" answer sheet has also been provided on page 46. This answer sheet is similar to those found on standardized tests. As your child completes each test, he or she can fill in the correct bubbles on the answer sheet.

How to Make the Most of This Book

Here are some useful ideas for optimizing the practice pages in this book.

- Set aside a specific place in your home to work on the practice pages. Keep it neat and tidy with materials on hand.

- Set up a certain time of day to work on the practice pages. This will establish consistency. An alternative is to look for times in your day or week that are less hectic and more conducive to practicing skills.

- Keep all practice sessions with your child positive and constructive. If the mood becomes tense or you and your child are frustrated, set the book aside and look for another time to practice with your child. Forcing your child to perform will not help. Do not use this book as a punishment.

- Help with instructions if necessary. If your child is having difficulty understanding what to do or how to get started, work the first problem through with him or her.

- Review the work your child has done. This serves as reinforcement and provides further practice.

- Allow your child to use whatever writing instruments he or she prefers. For example, colored pencils can add variety and pleasure to drill work.

- Pay attention to the areas in which your child has the most difficulty. Provide extra guidance and exercises in those areas. Allowing children to use drawings and manipulatives, such as coins, tiles, game markers, or flash cards, can help them grasp difficult concepts more easily.

- Look for ways to make real-life application to the skills being reinforced.

Practice 1

Solve the problems.

1. Timmy had 8 green marbles. He gave 6 marbles to his friend Mateo. How many marbles does Timmy have left?

   8 – 6 = ___

   He has _____ marbles left.

2. Jessie bought 2 stickers. Her father gave her some stickers. Now Jessie has 6 stickers in all. How many stickers did her father give her?

   2 + ___ = 6

   Her father gave her _____ stickers.

3. Uncle Jethro collects hats. He has 9 hats. He gave 7 to his nephew. How many hats does Uncle Jethro have left?

   9 – 7 = ___

   Uncle Jethro has ____ hats left.

4. Daisy has 8 pink bows and 1 yellow bow. How many bows does Daisy have in all?

   8 + 1 = ___

   She has _____ bows in all.

5. Suzy gathered 2 orange leaves. Then she gathered some brown leaves. She now has 7 leaves in all. How many brown leaves did she gather?

   2 + ___ = 7

   She gathered _____ brown leaves.

6. Brad picked several small flowers. He picked 2 large flowers. He now has 5 flowers. How many small flowers did Brad pick?

   2 + ___ = 5

   Brad picked _____ small flowers.

Practice 2 ⟋ ⟍ ⟋ ⟍ ⟋ ⟍ ⟋ ⟍ ⟋ ⟍ ⟋ ⟍ ⟋ ⟍

Solve the problems. Show your work.

1. My grandma made 4 fruitcakes. My mother made 5 fruitcakes. How many fruitcakes did they make in all?

   They made _____ fruitcakes in all.

2. My dad baked 5 hams. My grandpa baked 4 hams. How many hams did they bake in all?

   They baked _____ hams in all.

3. I made 8 cookies. My friend Mimi ate 2 cookies. How many cookies did we have left?

   We have _____ cookies left.

4. My mom cooked 7 chickens. My dad cooked 3 chickens. How many chickens did they cook in all?

   They cooked _____ chickens in all.

5. Hope picked 7 strawberries. Jim ate 4 of them. How many strawberries does Hope have left?

   Hope has _____ strawberries left.

6. Christie picked 3 lemons and 4 limes. How many lemons and limes does Christie have in all?

   Christie has _____ lemons and limes in all.

Practice 3

Solve the problems. Show your work.

1. Lassie ate 5 dog bones for breakfast and 9 for lunch. How many dog bones did Lassie eat in all?

   She ate _____ bones in all.

2. Benji barked 8 times at the mail carrier and 3 times at the meter reader. How many times did Benji bark?

   Benji barked _____ times in all.

3. Max bought 12 pounds of dog food. His dog Heidi ate 6 pounds of it. How many pounds of dog food does Max have left?

   Max has _____ pounds left.

4. The dog had 10 fleas. If 2 fleas jumped off, how many fleas were left?

   The dog had _____ fleas left.

5. Cybill had 11 dogs, but she gave 2 away. How many dogs does Cybill have now?

   Cybill has _____ dogs now.

6. Jesse's dog had 14 puppies, but 3 were given to his neighbor. How many puppies does Jesse have left?

   Jesse has _____ puppies left.

Practice 4

Solve the problems. Show your work.

1. Jenny made 8 kites. Tammy made 6 kites. How many kites did they make in all?

   They made _____ kites in all.

2. Chester saw 6 clouds that looked like lions and 9 clouds that looked like tigers. How many clouds did Chester see in all?

   Chester saw _____ clouds in all.

3. I have 13 books from my brother and sister. My sister gave me 7 of the books. How many books did my brother give me?

   My brother gave me _____ books.

4. Dad has 12 kites. If 7 of them are box kites, how many are not box kites?

   _____ are not box kites.

5. There were 9 dandelions in Terri's front yard and 7 in her backyard. How many dandelions were there in all?

   There were _____ dandelions in all.

6. Dennis earned 14 game points. He needs 18 points to win. How many more points must he earn to win?

   He must earn _____ more points.

Practice 5 ꙮ

Circle the correct number sentence.

1. Joan made 12 enchiladas. Her dog ate 6 of them. How many enchiladas were left?

   12 − 6 = 6   12 + 6 = 18

2. Hector ate 11 bags of nachos and 5 bowls of chili. How many things did Hector eat in all?

   11 + 5 = 16   11 − 5 = 6

3. Rachel made 7 tortillas. Her sister made 6 tortillas. How many tortillas were made in all?

   7 − 6 = 1   7 + 6 = 13

4. Hector made 10 enchiladas and 8 tacos. How many things did Hector make in all?

   10 + 8 = 18   10 − 8 = 2

5. Hope ate 15 fajitas. Grace ate 6 fewer fajitas than Hope. How many fajitas did Grace eat?

   15 + 6 = 21   15 − 6 = 9

6. Tony ate 16 peppers and 3 avocados. How many more peppers did Tony eat than avocados?

   16 + 3 = 19   16 − 3 = 13

Practice 6 ◗ ✺ ◗ ✺ ◗ ✺ ◗ ✺ ◗ ✺ ◗ ✺ ◗ ✺

Solve the problems. Show your work.

1. Penny has 2 red rings, 7 green rings, and 1 white ring. How many rings does Penny have in all?

Penny has _____ rings in all.

2. Christopher has 5 pennies, 2 nickels, and 7 dimes. How many coins does he have in all?

Christopher has _____ coins in all.

3. There were 13 aphids on Lucas' roses. He brushed 8 of them off. How many aphids were left?

There are _____ aphids left.

4. Jonas planted 18 flowers in his yard. The grasshoppers ate 9 of them. How many flowers does Jonas have left?

Jonas has _____ flowers left.

5. Doug collects flowers. He has 5 roses and 9 wildflowers in his collection. How many flowers does Doug have in all?

Doug has _____ flowers in all.

6. Trevor saw 13 cactuses. If 7 of them had flowers, how many cactuses did not have flowers?

_____ cactuses did not have flowers.

Practice 7 ৩ ৩ ৩ ৩ ৩ ৩ ৩ ৩ ৩ ৩ ৩ ৩ ৩ ৩ ৩

Solve the problems. Show your work.

1. Cory planted 13 carrot seeds. Only 5 sprouted. How many carrot seeds did not sprout?

   _____ carrot seeds did not sprout.

2. Raven planted 12 squash seeds and 6 tomato seeds. How many seeds did she plant in all?

   Raven planted _____ seeds in all.

3. Mr. Clover picked 16 heads of lettuce. He gave us 9 heads of lettuce. How many heads of lettuce does Mr. Clover have left?

   Mr. Clover has _____ heads of lettuce left.

4. Mrs. Peabody picked 15 ears of corn. If 6 of the ears had bugs, how many ears did not have bugs?

   _____ ears did not have bugs.

5. Cecil had 14 butterflies. If 8 of them flew away, how many butterflies does Cecil have left?

   Cecil has _____ butterflies left.

6. Kenya found 10 snails in her garden. She picked up 5 of them. How many snails are left in Kenya's garden?

   Kenya has _____ snails left in her garden.

Practice 8 ꙮ ꙮ ꙮ ꙮ ꙮ ꙮ ꙮ ꙮ ꙮ ꙮ ꙮ ꙮ ꙮ

Solve the problems. Show your work.

1. Ms. Smith put 15 vegetables into her pot of soup. If 7 of the vegetables were onions, how many were not onions?

_____ were not onions.

2. Mr. Wimple planted 18 radish seeds. The birds ate 8 of the seeds. How many radish seeds are left?

There are _____ radish seeds left.

3. A group of Girl Scouts went camping. If 13 of the girls set up camp while 5 went on a hike, how many Girl Scouts were there in all?

There were _____ Girl Scouts in all.

4. Eight Boy Scouts were busy cooking dinner. The rest were setting up the tents. There were 14 Boy Scouts in all. How many were setting up the tents?

_____ Boy Scouts were setting up the tents.

5. Liz bought 7 bird toys and 9 dog toys. How many animal toys did Liz buy in all?

Liz bought _____ toys in all.

6. Ana has 8 yellow fish and 7 silver fish. How many fish does Ana have in all?

Ana has _____ fish in all.

Practice 9 ⟳ ⟳ ⟳ ⟳ ⟳ ⟳ ⟳ ⟳ ⟳ ⟳ ⟳ ⟳ ⟳

Solve the problems. Show your work.

1. Dave picked up 8 blue oars and 9 gold oars. How many oars did Dave pick up in all?

   Dave picked up _____ oars in all.

2. Hilda gathered 6 small twigs and 9 large twigs to build a campfire. How many twigs in all did Hilda gather?

   Hilda gathered _____ twigs in all.

3. The Blue Troop was painting rocks. If 2 of the 18 members were painting rocks, how many members were not painting rocks?

   _____ were not painting rocks.

4. Nine of the Red Troop members were making belts. Five of the Orange Troop members were making belts, too. How many members were making belts?

   _____ were making belts.

5. The troops had 9 dozen cookies. They sold all of them. How many cookies do they have left?

   _____ cookies are left.

6. Mrs. Jones had 9 troops on her hike. She picked up 7 more troops along the way. How many troops does she now have?

   She has _____ troops now.

Practice 10

Solve the problems. Show your work.

1. The scouts were roasting marshmallows. They burned 3 of the 18 marshmallows. How many marshmallows did not burn?

_____ marshmallows did not burn.

2. At the crafts station, 6 scouts made bracelets and 8 made hats. How many crafts were made in all?

There were _____ crafts made in all.

3. The scouts spotted 5 bears and 9 cougars. How many animals did they see?

The scouts saw _____ animals.

4. One of the scouts sold 7 boxes of cookies on Monday and 6 boxes of cookies on Tuesday. How many boxes did she sell in all?

She sold _____ boxes in all.

5. Alvin found 16 cardboard boxes. He flattened 8 of them. How many of them were not flattened?

_____ boxes were not flattened.

6. Jamal had 15 spoons at the cooking station. If 3 of them fell off of the table, how many are left on the table?

There are _____ spoons left.

Adding Three Numbers

Practice 11

Read each word problem. In the box, write the number sentence it shows. Find the sum.

1.	2.

Kevin went for a walk and saw 1 frog, 3 cats, and 5 flowers. How many things did he see in all?	When Sally got on the school bus, there were 8 boys and 10 girls already there. How many children were there on the bus in all?
___ + ___ + ___ = ___	___ + ___ + ___ = ___

3.	4.

John ate a pizza with 7 mushrooms, 7 olives, and 5 pieces of pepperoni. How many toppings were on his pizza in all?	Today Jan saw 3 cats, 2 dogs, and 5 puppies in the park. How many animals did she see in all?
___ + ___ + ___ = ___	___ + ___ + ___ = ___

Practice 12 ✺ ✺ ✺ ✺ ✺ ✺ ✺ ✺ ✺ ✺ ✺ ✺

Solve the problems. Show your work.

1. Jacob has 5 books. Bart has 3 books. Amy has 2 books. How many books do they have in all?

They have _____ books in all.

2. The recipe calls for 1 cup of flour, 5 cups of sugar, and 1 cup of milk. How many cups of ingredients do I need to use?

I need to use _____ cups of ingredients.

3. Brittany collected 6 ladybugs, 2 grasshoppers, and 3 butterflies. How many insects did Brittany collect in all?

Brittany collected _____ insects in all.

4. There were 8 crows, 1 bluebird, and 1 robin sitting in a tree. How many birds were there in all?

There were _____ birds in all.

5. Rosita had 5 yellow pencils, 1 pink pencil, and 2 green pencils. How many pencils did she have in all?

Rosita had _____ pencils in all.

6. Luke had 4 baseball gloves, 5 bats, and 4 uniforms. How many pieces of baseball equipment did Luke have in all?

Luke had _____ pieces in all.

Practice 13

Solve the problems.

1. The elephant ate 50 large peanuts and 40 small peanuts. How many peanuts did the elephant eat?

$$
\begin{array}{r}
50 \\
+\ 40 \\
\hline
\end{array}
$$

The elephants ate _____ peanuts.

2. At the morning show, the lion jumped 70 feet. During the evening show, the lion jumped 25 feet. How far did the lion jump in all?

$$
\begin{array}{r}
70 \\
+\ 25 \\
\hline
\end{array}
$$

The lion jumped _____ feet in all.

3. There were 95 fleas in the flea circus. 60 fleas jumped away. How many fleas were left in the flea circus?

$$
\begin{array}{r}
95 \\
-\ 60 \\
\hline
\end{array}
$$

_____ fleas were left in the circus.

4. The monkeys ate 35 bananas in the morning and 20 bananas in the evening. How many bananas did the monkeys eat in all?

$$
\begin{array}{r}
35 \\
+\ 20 \\
\hline
\end{array}
$$

The monkeys ate _____ bananas in all.

5. The tiger has 85 stripes. The zebra has 30 fewer stripes than the tiger. How many stripes does the zebra have?

$$
\begin{array}{r}
85 \\
-\ 30 \\
\hline
\end{array}
$$

The zebra has _____ stripes.

6. The seal ate 85 pounds of fish on Monday and 55 fewer pounds of fish on Tuesday. How many pounds of fish did the seal eat on Tuesday?

$$
\begin{array}{r}
85 \\
-\ 55 \\
\hline
\end{array}
$$

The seal ate _____ pounds of fish on Tuesday.

Practice 14 ꙮ ꙮ ꙮ ꙮ ꙮ ꙮ ꙮ ꙮ ꙮ ꙮ ꙮ ꙮ ꙮ ꙮ

Solve the problems. Show your work.

1. Zipporah saw 15 paw prints. Lacey saw 13 fewer paw prints than Zipporah. How many paw prints did Lacey see?

   Lacey saw _____ paw prints.

2. Misty saw 43 turtles. Mabel saw 10 fewer turtles than Misty. How many turtles did Mabel see?

   Mabel saw _____ turtles.

3. Isaac has 22 peanuts. Pam has 17 more peanuts than Isaac. How many peanuts does Pam have?

   Pam has _____ peanuts.

4. Patty has 32 headbands. Cathy has 22 fewer headbands than Patty. How many headbands does Cathy have?

   Cathy has _____ headbands.

5. Candace went on the camel ride 41 times. Patrice went on the camel ride 18 times. How many times did the girls ride the camel in all?

   The girls rode the camel _____ times in all.

6. Sean fed the monkey 31 peanuts. Thomas fed the monkey 24 peanuts. How many peanuts did the monkey eat?

   The monkey ate _____ peanuts.

Adding and Subtracting 2-Digit Numbers Without Regrouping

Practice 15 ⟁ ⟁ ⟁ ⟁ ⟁ ⟁ ⟁ ⟁ ⟁ ⟁ ⟁ ⟁ ⟁

Solve the problems. Show your work.

1. There were 64 jelly beans in the jar. Now there are only 22 jelly beans. How many jelly beans are missing?

There are _____ jelly beans missing.

2. Mary Kaye invited 99 people to the party. Only 30 came. How many people did not come to the party?

_____ people did not come to the party.

3. Jeremy counted 96 stars on Monday and only 26 on Tuesday. How many fewer stars did Jeremy see on Tuesday than on Monday?

Jeremy saw _____ fewer stars on Tuesday.

4. Leo baked 30 cupcakes for his class. His dog ate 20 of the cupcakes. How many cupcakes does Leo have left?

Leo has _____ cupcakes left.

5. Hansel has 75 pieces of candy. He gives 35 pieces to Gretel. How many pieces of candy does Hansel have left?

Hansel has _____ pieces of candy left.

6. Ivan recycled 14 cans of soda and 25 bundles of newspapers. How many items did Ivan recycle in all?

Ivan recycled _____ items in all.

Practice 16

Read each word problem. In the box, write the number sentence it shows. Find the sum.

1.

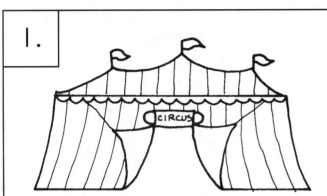

At the circus, Kenny saw 16 tigers and 27 monkeys. How many animals did he see in all?

_____ + _____ = _____

2.

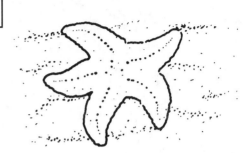

When Sandra went to the tidepools, she counted 28 starfish and 46 shells. How many things did she see in all?

_____ + _____ = _____

3.

During one month, Jared ate 27 sandwiches and 23 apples. How many things did he eat in all?

_____ + _____ = _____

4.

Emily did 19 addition problems and 33 subtraction problems at school. How many problems did she solve in all?

_____ + _____ = _____

Practice 17

Read each word problem. Write the number sentence it shows. Find the difference.

1.  Farmer Cole raised 93 bushels of wheat. Farmer Dale raised 68 bushels. What is the difference in the number of bushels each raised? _____ – _____ = _____	**2.**  Dennis scored 43 points in the basketball game. Claire scored 27. What is the difference in points each earned? _____ – _____ = _____
3.  Jason bought a pair of shoes for $53. Clark bought a pair for $28. What is the difference paid? $_____ – $_____ = $_____	**4.**  Jill counted 83 ants near an ant hill. Jack counted 65. What is the difference in the ants counted? _____ – _____ = _____

Practice 18 ㄱ ꝭ ꝯ ꝭ ꝯ ꝭ ꝯ ꝭ ꝯ ꝭ ꝯ ꝭ ꝯ ꝭ

Solve the problems. Show your work.

1. Derek had 61 balloons. If 19 of the balloons popped, how many balloons does Derek have left?

   Derek has _____ balloons left.

2. My sister Cheryl is 35 years old. I am 16 years old. How many years older is Cheryl?

   Cheryl is _____ years older.

3. Jerry unpacked the lightbulbs. There were 67. If 19 were broken, how many lightbulbs were not broken?

   There were _____ lightbulbs not broken.

4. Lupe needs 41 candles. There are only 23 candles. How many more candles does Lupe need?

   Lupe needs _____ more candles.

5. Eddie has 72 purple rings and 19 green rings. How many rings does Eddie have in all?

   Eddie has _____ rings in all.

6. Ariel once caught 38 insects in a net and 64 insects in a box. How many insects did Ariel catch in all?

   Ariel caught _____ insects in all.

Practice 19 ꙮ

Solve the problems. Show your work.

1. My cousin has 63 coins in his piggy bank. I have 21 coins in my piggy bank. How many coins do we have in all?

   We have _____ coins in all.

2. Becky is 23 years old. Her dad is 48 years old. If you added their ages together, how old would they be?

   They would be _____ years old.

3. When I was cleaning my room, I found 36 socks. I took 14 of them to the laundry room. How many socks do I still have in my room?

   I still have _____ socks in my room.

4. There were 71 birds in the bird show. Angelica saw only 46 of the birds. How many birds did Angelica not see?

   Angelica did not see _____ birds.

5. The students made 55 blueberry pancakes and 41 buttermilk pancakes. How many pancakes did they make in all?

   They made _____ pancakes in all.

6. The zookeeper had 86 crickets in a jar. She gave the tortoise 57 of the crickets. How many crickets does the zookeeper have left?

   The zookeeper has _____ crickets left.

Practice 20

When you add by groups, it is the same as multiplying. Look at the bags of marbles, and answer the questions.

1. How many bags?_____

   How many marbles in each bag?_____

   Altogether, how many marbles in all of the bags?_____

   We can write this as an addition equation like this:

   $$2 + 2 + 2 = 6$$

   Or, we can write it as a multiplication equation like this:

   $$3 \times 2 = 6$$

2. How many bags?_____

   How many marbles in each bag?_____

   Altogether, how many marbles in all of the bags?_____

   Write an addition equation for this problem.

   _____ + _____ = _____

   Write a multiplication equation for this problem.

   _____ x _____ = _____

Practice 21

Use multiplication to solve each word problem.

1. Layla played 3 rounds and won 2 marbles in each round.

$$3 \times 2 = \underline{\hspace{1cm}}$$

Layla won _____ marbles.

2. Mel played 2 rounds and won 4 marbles in each round.

$$2 \times 4 = \underline{\hspace{1cm}}$$

Mel won _____ marbles.

3. Cassie played 5 rounds and won 2 marbles in each round.

$$5 \times 2 = \underline{\hspace{1cm}}$$

Cassie won _____ marbles.

4. Mohammed played 1 round and won 2 marbles.

$$1 \times 2 = \underline{\hspace{1cm}}$$

Mohammed won _____ marbles.

5. Flavia played 6 rounds and won 2 marbles in each round.

$$6 \times 2 = \underline{\hspace{1cm}}$$

Flavia won _____ marbles.

6. Isidore played 2 rounds and won 7 marbles in each round.

$$2 \times 7 = \underline{\hspace{1cm}}$$

Isidore won _____ marbles.

Practice 22 ∂ ☙ ∂ ☙ ∂ ☙ ∂ ☙ ∂ ☙ ∂ ∂ ☙

Use multiplication to solve each word problem. Show your work.

1. Sebastian has 10 hats in 2 boxes. How many hats are there in all?

   There are _____ hats in all.

2. Eartha has 4 coin purses. Eartha has 3 coins in each purse. How many coins are there in all?

   There are _____ coins in all.

3. Santos has 5 flowers. Each flower has 2 buds. How many buds are there in all?

   There are _____ buds in all.

4. Maria had 3 books in 3 stacks. How many books are there in all?

   There are _____ books in all.

5. Tucker has 2 wheels on 4 bikes. How many wheels are there altogether?

   There are _____ wheels in altogether.

6. Gertie has 2 boxes. In each box there are 6 necklaces. How many necklaces are there in all?

   There are _____ necklaces in all.

Practice 23 ꙮ ꙮ ꙮ ꙮ ꙮ ꙮ ꙮ ꙮ ꙮ ꙮ ꙮ ꙮ ꙮ ꙮ ꙮ

Solve the problems. Show your work.

1. I had a whole pie. I cut it in half. How many pieces of pie do I have now?

   I have _____ pieces of pie.

2. Rocky ordered a pizza. The pizza was cut into 4 equal pieces. Rocky ate half of the pizza. How many pieces did Rocky eat?

   Rocky ate _____ pieces of pizza.

3. Ricky had 9 marbles. He kept 1/3. He gave 1/3 to Sonya and 1/3 to Len. How many marbles does each person now have?

   Each person has _____ marbles.

4. Marilyn had 5 houses. She sold 1/5 of them. How many houses did she keep?

   Marilyn kept _____ houses.

Practice 24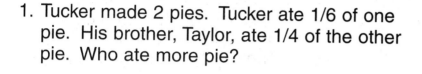

Shade the appropriate sections of the circle. Then solve the problem.

1. Tucker made 2 pies. Tucker ate 1/6 of one pie. His brother, Taylor, ate 1/4 of the other pie. Who ate more pie?

   _____ ate more pie.

$\frac{1}{6}$ $\frac{1}{4}$

2. Ivana ate 2/3 of a pizza. Donna ate 2/4 of a pizza. Who ate more pizza?

   _____ ate more pizza.

$\frac{2}{3}$ $\frac{2}{4}$

3. Sharon caught 4 out of 5 fly balls. Chad caught 4 out of 9 fly balls. Who caught more fly balls?

   _____ caught more fly balls.

$\frac{4}{5}$ $\frac{4}{9}$

4. Betty spelled 6 of the 7 words correctly. Les spelled 1 of the 6 words correctly. Which one earned the higher spelling score?

   _____ earned the higher score.

$\frac{6}{7}$ $\frac{1}{6}$

5. In basketball, Cori made 1 basket out of 8 tries. Danny made 1 basket out of 2 tries. Who had the higher shooting score?

   _____ had the higher shooting score.

$\frac{1}{8}$ $\frac{1}{2}$

Practice 25

Solve the problems.

1. The play began at 6:00 and lasted 45 minutes. What time did the play end?

   Show it on the clock.

2. Lee's family went on a picnic. They left at 11:00 A.M. and arrived 30 minutes later. What time did they get there?

   Show it on the clock.

   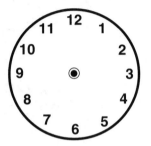

3. Luke does his homework one hour after dinner. If he eats dinner at 6:30 P.M., when does he start doing his homework?

   Circle the answer.

   5:30 P.M.   7:30 P.M.

4. Rachel eats lunch everyday at noon. When does Rachel eat lunch?

   Circle the answer.

   11:00 A.M.   12:00 P.M.

5. Susan comes home from school at 4:00 P.M. If she eats dinner two hours later, when does Susan eat dinner?

   Write the time.

   _____ : ___ ___

6. Sandy went to the Friday night school dance at 7:30 P.M. She arrived home 3 hours later. What time did Sandy get home?

   Write the time.

   _____ : ___ ___

Practice 26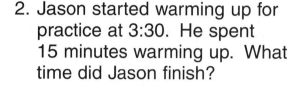

Circle the correct answer. Use the clock to help you.

1. It takes Shirley 1/2 an hour to walk to school. She left home at 8:00. What time did she arrive at school?

   half past 8   a quarter till 8

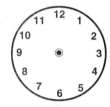

2. Jason started warming up for practice at 3:30. He spent 15 minutes warming up. What time did Jason finish?

   half past 3   a quarter till 4

3. Samantha spent 15 minutes raking leaves. She finished at 5:45. What time did Samantha start raking the leaves?

   half past 5   a quarter till 6

4. It took Brent 1/2 an hour to wax the car. He finished at 2:45. What time did Brent start waxing the car?

   half past 2   a quarter past 2

5. Shalom started cleaning her room at 7:00. She finished 15 minutes later. What time did Shalom finish cleaning her room?

   a quarter till 7   a quarter past 7

6. Roberto began cooking dinner at 4:15. It took him 1/2 an hour to finish cooking. What time was dinner on the table?

   half past 4   a quarter till 5

Practice 27 ♪ ☺ ♪ ☺ ♪ ☺ ♪ ☺ ♪ ☺ ♪ ♪ ☺

Solve the problems.

1. Sari spent 29¢ on gum and 30¢ on candy. How much money did she spend in all?

   Sari spent _____¢ in all.

2. Rob had 25¢. He earned 10¢ more by raking leaves. How much money does Rob have?

   Rob has _____¢ in all.

3. Ray had 45¢. He spent 15¢ on a toy. How much money does Ray have left?

   Ray has _____¢ left.

4. Deb had 95¢. She spent 85¢ on a toy for her cat. How much money does Deb have left?

   Deb has _____¢ left.

5. Bob had 35¢. He put 20¢ into his savings account. How much money does Bob have left?

   Bob has _____¢ left.

6. Sid had 90¢. He gave 60¢ to his brother. How much money does Sid have left?

   Sid has _____¢ left.

Practice 28 ⟡ ⟡ ⟡ ⟡ ⟡ ⟡ ⟡ ⟡ ⟡ ⟡ ⟡ ⟡ ⟡

Solve the problems.

1. John has 3 dimes. How much money does he have? Does he have enough to buy a 25¢ candy?

   John has _____¢.

   Yes   No

2. Franklin has a quarter, a dime, and a penny. How much money does he have in all? Does he have enough to buy a 50¢ cookie?

   Franklin has _____¢.

   Yes   No

3. I have one quarter. How many nickels does it take to make one quarter?

   It takes _____ nickels to make one quarter.

4. I have one quarter. How many pennies does it take to make one quarter?

   It takes _____ pennies to make one quarter.

5. Joe had 34¢ in his pocket. If 21¢ fell out of the hole in his pocket, how much money does he have left?

   Joe has _____¢ left.

6. Esmerelda had 95¢ in her lunch bag. She gave her friend 14¢. How much money does she have left in her lunch bag?

   Esmerelda has _____¢ left in her lunch bag.

Practice 29 ⟋ ⟋ ⟋ ⟋ ⟋ ⟋ ⟋ ⟋ ⟋ ⟋ ⟋ ⟋ ⟋ ⟋ ⟋

Solve the problems. Show your work.

1. Helen earned 75¢ washing the car. She put 50¢ in her bank. How much money does Helen have left?

Helen has _____¢ left.

2. Val's grandma gave her 95¢. Val spent 50¢ on a book. How much money does she have left?

Val has _____¢ left.

3. Christine cleaned the pool for 85¢. She bought a pool toy for 73¢. How much money does Christine have left?

Christine has _____¢ left.

4. Mom had 85¢ in her wallet. She spent 50¢ on a carton of juice. How much money does she have left?

Mom has _____¢ left.

5. Mia spent 55¢ on popcorn and 20¢ for gum. How much money did Mia spend in all?

Mia spent _____¢ in all.

6. Elvis bought an eraser for 45¢ and a pencil for 20¢. How much money did Elvis spend in all?

Elvis spent _____¢ in all.

Practice 30 ৹ ৹ ৹ ৹ ৹ ৹ ৹ ৹ ৹ ৹ ৹ ৹ ৹ ৹

Solve the problems.

1. Macy has 4 coins in her pocket. Together they make 8¢. What coins does Macy have in her pocket?

Macy has _____

_____.

2. Joel has 3 coins in his pocket. Together the coins make 20¢. What coins does Joel have in his pocket?

Joel has _____

_____.

3. Cora has 5 coins in her pocket. Together the coins make 25¢. What coins does Cora have in her pocket?

Cora has _____

_____.

4. Lane has 6 coins in his pocket. Together the coins make 32¢. Two of the coins are dimes. What are the other 4 coins?

Lane has _____

_____.

5. Elita has 6 coins in her pocket. Together the coins make 23¢. Two of the coins are nickels. What are the other 4 coins?

Elita has _____

_____.

6. Rex has 2 coins in his pocket. Together the coins make 11¢. What coins does Rex have in his pocket?

Rex has _____

_____.

Practice 31

Write the subtraction problem and solve.

$42.16	$51.37	$10.48	$30.66	$24.79	$11.06

1. Rosa has $52.20. She buys an area rug.
   How much money does she have left?

   $52.20
   − $42.16
   _____

   Rosa has _____ left.

2. Theo has $79.50. He buys a new couch.
   How much money does Theo have left?

   $__ __.__ __
   − $__ __.__ __
   _____
   $__ __.__ __

   Theo has _____ left.

3. Thomas has $11.75. He buys a new lamp.
   How much money does Thomas have left?

   $__ __.__ __
   − $__ __.__ __
   _____
   $__ __.__ __

   Thomas has _____ left.

4. Maurine has $38.95. She buys a new TV set.
   How much money does Maurine have left?

   $__ __.__ __
   − $__ __.__ __
   _____
   $__ __.__ __

   Maurine has _____ left.

5. Monty has $86.93. He buys a new table.
   How much money does Monty have left?

   $__ __.__ __
   − $__ __.__ __
   _____
   $__ __.__ __

   Monty has _____ left.

6. Susan has $13.10. She buys a chair. How
   much money does Susan have left?

   $__ __.__ __
   − $__ __.__ __
   _____
   $__ __.__ __

   Susan has _____ left.

Practice 32 ꙮ ꙮ ꙮ ꙮ ꙮ ꙮ ꙮ ꙮ ꙮ ꙮ ꙮ ꙮ ꙮ ꙮ

Fill in the answer circle to each problem.

1. Richard is thinking of a number.
   The number has a 9 in the ones place
   and a 4 in the hundreds place.
   What is Richard's number?

934	349	409
◯	◯	◯

2. Maggie is thinking of a number.
   The number has an 8 in the tens place
   and an 8 in the hundreds place.
   What is Maggie's number?

883	838	388
◯	◯	◯

3. Jack is thinking of a number.
   The number has a 9 in the tens place
   and a 1 in the ones place.
   What is Jack's number?

219	291	912
◯	◯	◯

4. Celia is thinking of a number.
   The number has a 5 in the hundreds
   place and a 4 in the tens place.
   What is Celia's number?

745	457	547
◯	◯	◯

5. Alec is thinking of a number.
   The number has a 2 in the tens place
   and a 2 in the ones place.
   What is Alec's number?

262	622	226
◯	◯	◯

Finding Greater Than or Less Than

Practice 33

Use the > (greater than) or < (less than) symbol. Then solve the problem.

1. Marvin has 659 tickets. Michelle has 337 tickets. Who has more tickets?

   659 ◯ 337

   _____ has more tickets.

2. Ellen has 120 cards. Edith has 562 cards. Who has more cards?

   120 ◯ 562

   _____ has more cards.

3. Peter caught 261 butterflies. Paula caught 892 butterflies. Who caught more butterflies?

   261 ◯ 892

   _____ caught more butterflies.

4. Sally counted 443 birds. Steven counted 434 birds. Who counted more birds?

   443 ◯ 434

   _____ counted more birds.

5. Ron found 516 pennies. Barb found 798 pennies. Who found more pennies?

   516 ◯ 798

   _____ found more pennies.

6. Rebecca sewed on 185 buttons. Heather sewed on 981 buttons. Who sewed on more buttons?

   185 ◯ 981

   _____ sewed on more buttons.

7. Mark made 429 baskets. Maria made 451 baskets. Who made more baskets?

   429 ◯ 451

   _____ made more baskets.

8. Zina collected 382 seashells. Zack collected 665 seashells. Who collected more seashells?

   382 ◯ 665

   _____ collected more seashells.

Practice 34 ꙮ ꙮ ꙮ ꙮ ꙮ ꙮ ꙮ ꙮ ꙮ ꙮ ꙮ ꙮ ꙮ ꙮ

Fill in the circle of the correct answer.

1. Jeff has $4 to buy 2 bottles of ketchup for the class picnic. The ketchup costs $1.20 a bottle. How much money will Jeff have left after he buys the ketchup?

   ◯ $1.20   ◯ $2.40
   ◯ $1.60   ◯ $2.60

2. Della has 22 sports cards. She has 7 baseball cards and 8 football cards. The rest of the cards are soccer cards. How many soccer cards does Della have?

   ◯ 7   ◯ 8
   ◯ 14   ◯ 15

3. Chan made 17 origami frogs. He gave 5 to Lin and 3 to Maria. How many frogs did Chan have left?

   ◯ 3   ◯ 5
   ◯ 8   ◯ 9

4. Gary's allowance is $5 a week, and Shamika's is $4 a week. Over a period of 4 weeks, how much more money does Gary receive?

   ◯ $5   ◯ $3
   ◯ $4   ◯ $2

5. Marilyn and her family went on a vacation for 14 days. They spent 5 days in Germany and 4 days in Italy. They spent the rest of the time in France. How many days did they spend in France?

   ◯ 10   ◯ 5
   ◯ 9   ◯ 4

6. Javier saved $50 to buy new clothes for school. He bought a shirt for $14 and a pair of jeans for $12. How much money does Javier have left for shoes?

   ◯ $24   ◯ $38
   ◯ $36   ◯ $20

Practice 35

Read the clues. Cross out the information that is not needed to solve the problem. Write the student's house number on the line.

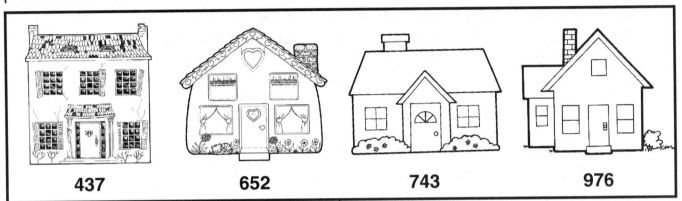

437   **652**   **743**   **976**

Roy's House

- Roy lives in the brown house.

- The numbers 3 and 7 are in his house address.

- His house has the lowest number.

 Roy lives in house number _____.

Sue's House

- Sue's house does not have the highest number on the street.

- Sue's house is the prettiest.

- Sue's house has a 6 in the address.

 Sue lives in house number _____.

Bob's House

- Bob's house number has a 7 in it.

- Bob's house does not have the highest address on the street.

- Bob really likes his house.

 Bob lives in house number _____.

Marie's House

- Marie's house does not have a 3 as part of the address.

- Marie's house is really big.

- Marie's house has a 7 in the tens place.

 Marie lives in house number _____.

Practice 36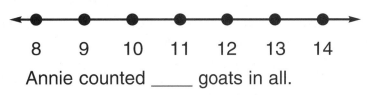

Use the number line to help find the answer.

1. Marybeth counted 9 goats. Annie counted 5 more goats than Marybeth. How many goats did Annie count in all?

   ← • — • — • — • — • — • — • →
   8   9   10   11   12   13   14

   Annie counted _____ goats in all.

2. Gordon counted 11 red cows and 4 brown cows. How many cows did Gordon count in all?

   ← • — • — • — • — • — • — • →
   10   11   12   13   14   15   16

   Gordon counted _____ cows in all.

3. Amos counted 10 ewes and 6 lambs. How many sheep did Amos count in all?

   ← • — • — • — • — • — • — • →
   10   11   12   13   14   15   16

   Amos counted _____ sheep in all.

4. Sheila had 15 pigs. She sold 4 of them at the state fair. How many pigs does Sheila have left?

   ← • — • — • — • — • — • — • →
   10   11   12   13   14   15   16

   Sheila has _____ pigs left.

5. Rob had 17 ponies. 3 of the ponies were brown. The rest were black. How many black ponies did Rob have?

   ← • — • — • — • — • — • — • →
   12   13   14   15   16   17   18

   Rob had _____ black ponies.

6. Daisy has 16 goats. Hazel has 2 fewer goats than Daisy. How many goats does Hazel have?

   ← • — • — • — • — • — • — • →
   10   11   12   13   14   15   16

   Hazel has _____ goats.

Test Practice 1

Directions: Fill in the circle of the correct anwer.

Sample

Mrs. Taylor planted 6 pink rose bushes and 5 white rose bushes in her front yard. Which number sentence tells how to find the number of rose bushes she planted altogether?

(A) $5 + 1 = 6$ (C) $6 - 5 = 1$

(B) $6 - 1 = 5$ (D) $6 + 5 = 11$

---

1. Maria bought 5 candy bars on Monday. On Wednesday she bought 3 more. How many candy bars did Maria buy in all?

   (A) 2   (C) 3

   (B) 5   (D) 8

2. Ron has 5 baseball cards, 7 football cards, and 2 soccer cards. How many sports cards does Ron have in all?

   (A) 7   (C) 9

   (B) 12   (D) 14

3. Lin and José were playing checkers. Lin won 6 games and José won 4 games. How many games did they play?

   (A) 10   (C) 6

   (B) 4   (D) 2

4. At the Springton Zoo Reptile House there are 11 lizards, 25 snakes, and 7 turtles. How many reptiles are there in all?

   (A) 34   (C) 43

   (B) 36   (D) 32

5. Gus bought 2 tickets to the school play. The tickets cost $4 each. Which number sentence tells how to find out how much money Gus spent altogether?

   (A) $\$2 + \$4 = \$6$ (C) $\$4 + \$4 = \$8$

   (B) $\$6 - \$4 = \$2$ (D) $\$8 - \$4 = \$4$

6. In the morning there were 7 crows on the old oak tree. In the afternoon 8 more crows joined them. How many crows in all were in the old oak tree?

   (A) 7   (C) 8

   (B) 15   (D) 16

Test Practice 2 ꜟ ꙮ ꜟ ꙮ ꜟ ꙮ ꜟ ꙮ ꜟ ꙮ ꜟ ꜟ ꙮ

Directions: Fill in the circle of the correct anwer.

Sample

On the east side of Oak Tree Lane there are 12 houses. There are 9 houses on the west side. How many more houses are there on the east side?

(A) 12

(B) 3

(C) 9

(D) 5

---

1. On Sunday, 35 children were playing in the park. Then 14 of them went home. How many children were still playing?

(A) 21

(B) 11

(C) 49

(D) 14

---

2. Jesse bought a bat for $13. He paid for it with a $20 bill. How much change did he get back?

(A) $4

(B) $6

(C) $5

(D) $7

---

3. Fred colored 24 Easter eggs. He colored 12 of them yellow and 12 of them blue. Fred gave 18 of them away to his friends. Which number sentence tells how to find the number of Easter eggs he had left?

(A) $24 + 18 = 42$

(B) $24 - 12 = 12$

(C) $24 - 18 = 6$

(D) $18 - 12 = 6$

---

4. Martha set the table for 9 people. Only 7 people were able to come. How many sets of silverware should Martha take off of the table?

(A) 2

(B) 5

(C) 7

(D) 9

---

5. A group of 12 people planned to pick up their tickets at the box office on the night of the big game. Only 8 of these people made it to the game. How many tickets were left at the box office?

(A) 12

(B) 5

(C) 8

(D) 4

---

6. There were 55 students who tried out for the school play. Only 33 students got parts. Which number sentence tells how to find out how many students did not get parts?

(A) $55 + 33 = 88$

(B) $33 + 33 = 66$

(C) $55 - 33 = 22$

(D) $55 - 10 = 45$

Test Practice 3 ꙮ ꙮ ꙮ ꙮ ꙮ ꙮ ꙮ ꙮ ꙮ ꙮ ꙮ

Directions: Do each multiplication problem. Fill in the circle of the correct answer.

Sample

Mr. Wilson gave 2 candy bars each to Franklin, Pham, and José. How many candy bars did he give them in all?

(A) 2
(B) 3
(C) 6
(D) 12

Franklin

Pham

José

---

1. Ms. Anderson, the school secretary, sharpened some new pencils. She put 4 pencils into each of 3 containers. She put one container on the principal's desk, one on the school nurse's desk, and one on her own desk. Which picture shows the right number of pencils and containers?

(A)

(B)

(C)

(D)

---

2. It takes 4 cups of punch to fill a pitcher. Betsy wants to fill 3 pitchers with punch. How many cups of punch will she need?

    =

(A) 4 cups
(B) 10 cups
(C) 8 cups
(D) 12 cups

---

3. Lupe and Kelly each decorated 6 eggs. They want to display them together. Which container will hold all of their eggs?

(A)

(B)

(C)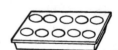

(D)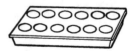

Test Practice 4 ⟳ ⟳ ⟳ ⟳ ⟳ ⟳ ⟳ ⟳ ⟳ ⟳ ⟳ ⟳

Directions: Fill in the correct answer.

1. Mark had 9 toy cars. He kept 1/3 of them. He gave 1/3 to Tony and 1/3 to Ben. How many toy cars does each person now have?

   (A) 1   (B) 2   (C) 3   (D) 4

2. Jonathan had 5 CDs. He sold 1/5 of them. How many CDs did he keep?

   (A) 4   (B) 3   (C) 2   (D) 1

3. Loren comes home from work at 5:00 P.M. If she eats dinner two hours later, when does Loren eat dinner?

   (A) 5:30 P.M.   (B) 6:00 P.M.   (C) 6:30 P.M.   (D) 7:00 P.M.

4. Steve went to the Saturday night movie at 7:30 P.M. He arrived home 3 hours later. What time did Steve get home?

   (A) 9:30 P.M.   (B) 10:00 P.M.   (C) 10:30 P.M.   (D) 11:00 P.M.

5. Dana started soccer practice at 2:30 P.M. and finished 15 minutes later. What time did Dana finish?

   (A) a quarter past 2   (B) a quarter till 2   (C) a quarter till 3   (D) 3:00

6. Bob practices the guitar every day for 1/2 an hour. If Bob needs to finish by 4:45, what is the latest time he can begin practicing the guitar?

   (A) half past 4   (B) a quarter past 4   (C) half past 3   (D) a quarter past 3

Test Practice 5

Directions: Fill in the correct answer.

1. Sara spent 63¢ on a doll and 23¢ for a pencil. How much money did Sara spend in all?

   Ⓐ 40¢　　　　Ⓑ 80¢　　　　Ⓒ 68¢　　　　Ⓓ 86¢

2. Jeff had 75¢. He bought a candy bar for 54¢. How much money did Jeff have left?

   Ⓐ 25¢　　　　Ⓑ 21¢　　　　Ⓒ 12¢　　　　Ⓓ 29¢

3. James has 2 coins in his pocket. Together the coins make 15¢. What coins does James have in his pocket?

   Ⓐ　　　　　　Ⓑ　　　　　　Ⓒ　　　　　　Ⓓ

4. Barb has 3 coins in her pocket. Together the coins make 25¢. What coins does Barb have in her pocket?

   Ⓐ　　　　　　Ⓑ　　　　　　Ⓒ　　　　　　Ⓓ

5. Peter has $76.43. He buys a new table that cost $45.31. How much money does Peter have left?

   Ⓐ $121.74　　Ⓑ $31.74　　Ⓒ $31.12　　Ⓓ $31.21

6. Joyce has $15.12. She buys a chair that cost $11.06. How much money does Joyce have left?

   Ⓐ $4.06　　　Ⓑ $26.18　　Ⓒ $4.18　　　Ⓓ $4.60

Test Practice 6

Directions: Fill in the correct answer.

1. Joanna is thinking of a number. The number has a 7 in the hundreds place and a 4 in the tens place. What is Joanna's number?

   (A) 745   (B) 457   (C) 547   (D) 574

2. Ralph is thinking of a number. The number has a 3 in the tens place and a 3 in the ones place. What is Ralph's number?

   (A) 63   (B) 633   (C) 336   (D) 636

3. Cross out the unimportant information and solve.
   Jade has 6 friends. Each friend has 2 cookies. The friends like the cookies. How many cookies did they have in all?

   (A) 11   (B) 12   (C) 8   (D) 4

4. Cross out the unimportant information and solve.
   Jacob had 20 masks. He sold 12 of them at the fair. His favorite mask is green. How many masks are left?

   (A) 6   (B) 32   (C) 12   (D) 8

5. Lee counted 8 candies. Eric counted 5 more candies than Lee. How many candies did Eric count in all?

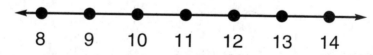

   8   9   10   11   12   13   14

   (A) 12   (B) 13   (C) 11   (D) 14

6. Paul counted 11 red cars and 4 brown cars. How many cars did Paul count in all?

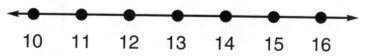

   10   11   12   13   14   15   16

   (A) 13   (B) 14   (C) 16   (D) 15

Answer Sheet

Test Practice 1 (page 40)	Test Practice 2 (page 41)	Test Practice 3 (page 42)
Sample (A) (B) (C) (D)	**Sample** (A) (B) (C) (D)	**Sample** (A) (B) (C) (D)
1. (A) (B) (C) (D) 2. (A) (B) (C) (D) 3. (A) (B) (C) (D) 4. (A) (B) (C) (D) 5. (A) (B) (C) (D) 6. (A) (B) (C) (D)	1. (A) (B) (C) (D) 2. (A) (B) (C) (D) 3. (A) (B) (C) (D) 4. (A) (B) (C) (D) 5. (A) (B) (C) (D) 6. (A) (B) (C) (D)	1. (A) (B) (C) (D) 2. (A) (B) (C) (D) 3. (A) (B) (C) (D)

Test Practice 4 (page 43)	Test Practice 5 (page 44)	Test Practice 6 (page 45)
1. (A) (B) (C) (D) 2. (A) (B) (C) (D) 3. (A) (B) (C) (D) 4. (A) (B) (C) (D) 5. (A) (B) (C) (D) 6. (A) (B) (C) (D)	1. (A) (B) (C) (D) 2. (A) (B) (C) (D) 3. (A) (B) (C) (D) 4. (A) (B) (C) (D) 5. (A) (B) (C) (D) 6. (A) (B) (C) (D)	1. (A) (B) (C) (D) 2. (A) (B) (C) (D) 3. (A) (B) (C) (D) 4. (A) (B) (C) (D) 5. (A) (B) (C) (D) 6. (A) (B) (C) (D)

Answer Key

Page 4
1. 2
2. 4
3. 2
4. 9
5. 5
6. 3

Page 5
1. 9
2. 9
3. 6
4. 10
5. 3
6. 7

Page 6
1. 14
2. 11
3. 6
4. 8
5. 9
6. 11

Page 7
1. 14
2. 15
3. 6
4. 5
5. 16
6. 4

Page 8
1. 12 − 6 = 6
2. 11 + 5 = 16
3. 7 + 6 = 13
4. 10 + 8 = 18
5. 15 − 6 = 9
6. 16 − 3 = 13

Page 9
1. 10
2. 14
3. 5
4. 9
5. 14
6. 6

Page 10
1. 8
2. 18
3. 7
4. 9
5. 6
6. 5

Page 11
1. 8
2. 10
3. 18
4. 6
5. 16
6. 15

Page 12
1. 17
2. 15
3. 16
4. 14
5. 0
6. 16

Page 13
1. 15
2. 14
3. 14
4. 13
5. 8
6. 12

Page 14
1. 1 + 3 + 5 = 9
2. 1 + 8 + 10 = 19
3. 7 + 7 + 5 = 19
4. 3 + 2 + 5 = 10

Page 15
1. 10
2. 7
3. 11
4. 10
5. 8
6. 13

Page 16
1. 90
2. 95
3. 35
4. 55
5. 55
6. 30

Page 17
1. 2
2. 33
3. 39
4. 10
5. 59
6. 55

Page 18
1. 42
2. 69
3. 70
4. 10
5. 40
6. 39

Page 19
1. 16 + 27 = 43
2. 28 + 46 = 74
3. 27 + 23 = 50
4. 19 + 33 = 52

Page 20
1. 93 − 68 = 25
2. 43 − 27 = 16
3. $53 − $28 = $25
4. 83 − 65 = 18

Page 21
1. 42
2. 19
3. 48
4. 18
5. 91
6. 102

Page 22
1. 84
2. 71
3. 22
4. 25
5. 96
6. 29

Page 23
1. 3, 2, 6
2. 2, 4, 8, 4 + 4 = 8, 4 x 2 = 8

Page 24
1. 6
2. 8
3. 10
4. 2
5. 12
6. 14

Page 25
1. 20
2. 12
3. 10
4. 9
5. 8
6. 12

Page 26
1. 2
2. 2
3. 3
4. 4

Page 27
1. Taylor
2. Ivana
3. Sharon
4. Betty
5. Danny

Answer Key (cont.)

Page 28

1.

2. (clock)

3. 7:30 P.M.
4. 12:00 P.M.
5. 6:00 P.M.
6. 10:30 P.M.

Page 29

1. half past 8
2. a quarter till 4
3. half past 5
4. a quarter past 2
5. a quarter past 7
6. a quarter till 5

Page 30

1. 59¢
2. 35¢
3. 30¢
4. 10¢
5. 15¢
6. 30¢

Page 31

1. 30¢, Yes
2. 36¢, No
3. 5
4. 25
5. 13¢
6. 81¢

Page 32

1. 25¢
2. 45¢
3. 12¢
4. 35¢
5. 75¢
6. 65¢

Page 33

1. 1 nickel, 3 pennies
2. 2 nickels, 1 dime
3. 5 nickels
4. 2 nickels, 2 pennies
5. 1 dime, 3 pennies
6. 1 penny, 1 dime

Page 34

1. $52.20 − $42.16 = $10.04
2. $79.50 − $51.37 = $28.13
3. $11.75 − $10.48 = $1.27
4. $38.95 − $30.66 = $8.29
5. $86.93 − $24.79 = $62.14
6. $13.10 − $11.06 = $2.04

Page 35

1. 409
2. 883
3. 291
4. 547
5. 622

Page 36

1. >, Marvin
2. <, Edith
3. <, Paula
4. >, Sally
5. <, Barb
6. <, Heather
7. <, Maria
8. <, Zack

Page 37

1. $1.60
2. 7
3. 9
4. $4
5. 5
6. $24

Page 38

Roy's House: 437
  Cross out: Roy lives in the brown house.
Sue's House: 652
  Cross out: Sue's house is the prettiest.
Bob's House: 743
  Cross out: Bob really likes his house.
Marie's House: 976
  Cross out: Marie's house is really big.

Page 39

1. 14
2. 15
3. 16
4. 11
5. 14
6. 14

Page 40

Sample: D
1. D
2. D
3. A
4. C
5. C
6. B

Page 41

Sample: B
1. A 2. D 3. C 4. A 5. D 6. C

Page 42

Sample: C
1. B 2. D 3. D

Page 43

1. C 2. A 3. D 4. C 5. C 6. B

Page 44

1. D
2. B
3. C
4. A
5. C
6. A

Page 45

1. A 2. B
3. B—Cross out: The friends like the cookies.
4. D—Cross out: His favorite mask is green.
5. B 6. D